The Genealogy of the Locomobile Steam Carriage 1899 - 1904

The Genealogy of the

Locomobile Steam Carriage

1899 - 1904

by Donald L. Ball

Published by

The Stanley Museum, Inc.

Kingfield, Maine, USA

1994

First Edition

Published in the United States of America by
The Stanley Museum, Inc.
School Street, Post Office Box 280
Kingfield , Maine 04947 USA
Telephone 207-265-2729—Facsimile 207-265-4700

Printed on Warren Lustro Dull in Times Roman, cover stock Marine Mohawk Vellum
by Carrabassett Printers, North Anson, Maine, USA

Photo Credits: All photographs by Donald L. Ball, with the exception of the following:
Jay Simpson: page 3; top of page 8; top of page 18; bottom of page 20, and top of page 29
Ron Thurber: top two on page 10; bottom on page 29 and 30
Original factory photographs courtesy Mark Herman: cover, frontispiece and pages 9 and 39
Special thanks to Richard French for the use of his pristine,
original 1900 Locomobile Catalogue for the Appendix

Cover Photo: Circa 1902 Locomobile Wagonette, in front of the factory in Bridgeport, Connecticut,
original factory photographs courtesy Mark Herman.

The Genealogy of the

Locomobile

Steam Carriage

Contents

Foreword

It seems very appropriate that the Stanley Museum publish this compilation of information about a little-known and important part of automotive history. It is in keeping with the mission of the Museum to preserve, interpret and promote the early history of this century so intimately tied to the many accomplishments of the remarkable Stanley family. None is more closely associated than Locomobile, as Don Ball points out on page 1.

If "Stanley" is followed by "Steamer" as part of the national idiom, Locomobile is hot on its heels as a premier example of steam automotive technology.

The Stanley Museum, founded in 1981, has come to serve an important and very particular segment of the automotive history field — steam. Indeed, because this Museum is not a transportation museum but rather a history museum, we are in the unique position of being the only museum in the world able to concentrate on an automotive technology that otherwise gathers dust in the corner of large, universal collections.

It is therefore with particular pleasure that we participate in shedding light on yet another little-known — and important — part of history. May this be one of many such publications in our future.

Susan S. Davis
Executive Director
Stanley Museum, Inc.
Kingfield, Maine, USA

September 1994

Acknowledgment

When I started the restoration of my 1901 Locomobile, I little realized how much research I would have to do just to find out what was right for my car. For an automobile that was manufactured for just five years, they made a disproportionate number of changes in the various parts making up the car.

As my restoration project progressed, not only did I amass information on my particular car but I found a considerable amount of data on all Locomobiles. After my car was completed, an inquiry from another asking for information on the Locomobile prompted me to set down what I knew in writing. Showing the manuscript to Sue Davis of the Stanley Museum led her to ask to publish it in book form. I now submit this for the approval of the reader.

I do not claim to know all there is to know — one individual cannot amass all the required information himself — and would be very interested to hear of any errors or incomplete data. There's always time for an expanded edition.

This is a project that has had the assistance of many in the antique car hobby. I would particularly like to thank Mike Clair, my partner in research as we restored our respective Locomobile project cars. Art Hart of Vintage Steam Products added bits of information based on his restoration of a 1902 Locomobile and proofread the manuscript for accuracy.

Thanks also go to Jay Simpson, Tom Wozniak, Eric Wagner and especially Garyl Turley and Herb Ottaway who took the time to read the manuscript and offer suggestions.

Don Ball
Santa Clarita, CA

May 1994

"*Locomobile*"

For anyone who has attempted the accurate restoration of a Locomobile steam car or any other early vehicle, the problem of accurately dating a car and determining the proper engine and other accessories can be overwhelming.

Although many cars have had reams of documentation with many indications of which models were provided with which equipment, the early Locomobiles have had practically none. The only attempt of which I am aware was Chris Figge's article entitled "The Chicken Coop Special" which appeared in the May-June, 1977 issue of the *Horseless Carriage Gazette*. In this article, Mr. Figge was primarily interested in his early 1899-model steamer but did list various features for the different model years.

L.J. Andrew Villalon wrote three articles for *The Bulb Horn* in the April/June, 1981 and the July/Sept and October/December, 1986 issues. Although these articles were concerned with the business history of the Locomobile Company, they did provided several clues that help the dating of these cars.

Before going further, I have to add that my conclusions do not represent all there is to know about these cars but are simply my opinions based on the information following. It contradicts some information already extant but if anyone has better information, I would very much like to hear from him or her so as to get the story as accurate as possible.

A brief history of Locomobile

F.E. Stanley built three small steam cars in 1897 and 1898 and with his twin brother F.O. exhibited one of them at a public horseless carriage show at Charles River Park in November, 1898. Due to a fantastic public response, they decided to start producing these cars commercially. Before this could actually begin, however, John Brisben Walker, owner of *Cosmopolitan* magazine, and Amzi Lorenzo Barber, the nation's leading producer of asphalt, purchased the rights to the Stanley car. Almost immediately, these two partners separated, each having the rights to produce the Stanley car. Walker went down to New York to start work on what was to become the Mobile steamer while Barber remained in Watertown, Massachusetts, to produce the Locomobile.

Serial Numbers

The first car model or style made by Locomobile was Style 1, quickly superseded by the Style 2 which became the "standard" model made by the company. The features of the Style 1 were well-described in Figge's article and this car has been traditionally referred to as having been made in 1899. I believe that very few cars were actually made in 1899, some of which were the larger, sturdier Style 2, and that the Style 1 was made concurrently with it.

When the Stanleys decided to get into the steam car business, they ordered enough parts to make about 200 cars, all of which went to Barber when the Stanleys sold the company. These parts,

I believe, were all for what became the Style 1 car. In December, 1899, Locomobile started advertising the Style 2 and other styles. It is my feeling that Barber immediately, upon buying the company and the parts for 200 cars, realized the Style 1 was too light and began to design a heavier car. He was a businessman, however, and was going to sell those 200 cars.

The Company was incorporated in June, 1899, and established the name of Locomobile on July 29, 1899. The first ads appeared in August, 1899. An October, 1899, ad proclaims "Delivery in 90 Days," implying that the cars weren't quite finished yet.

Patents had been applied for by the Stanleys for several features of the car, the first of which were granted on November 14, 1899. The earliest original serial number plate I have seen was Number 221 and it was on a plate with the November 14, 1899 date cast into it. What does this mean? The serial number plate obviously was made no earlier than November 14, 1899 since the patents weren't issued until then. Owing to normal communication delays and getting the plates cast, this particular car probably could not have come out of the factory before the first of December, 1899.

In February, 1900, an ad appeared indicating that there were "500 cars on the road." When considering that this probably included some typical salesman's puffery, it tends to indicate that very few cars were produced in 1899. A similar ad in August, 1900 stated "1500 Locomobiles are in actual use."

Based on the total number of cars produced, Locomobile made an average of about 160-200 cars per month. Another original car, Serial No. 2258, has a nameplate where the last patent date of December 11, 1900 is stamped, not cast, onto the plate. Since it is stamped on the plate, it can be assumed that Locomobile was concerned about keeping track of its latest patent. The next patent the company obtained was February 18, 1901. Therefore, this particular car was evidently made between December 11, 1900 and February 18, 1901. This supports the average production figure arrived at above.

By June, 1901, Locomobile claimed to have produced 3000 cars. This again tracks relatively well with the monthly estimate. Based on all of this, I have developed an annual serial number list which I think is fairly accurate.

1899	Serial Numbers 1-350
1900	Serial Numbers 351-2400
1901	Serial Numbers 2401-4660
1902	Serial Numbers 4661-?

Beyond 1902, I have not seen any original cars that would substantiate any reasonable conclusions.

It should be noted that the serial number of the engine corresponded to the serial number of the car except for the first few cars using the Mason engine. In other words, an engine with serial number "1234" came from the car with the serial number "1234" stamped on the nameplate.

Engines

Most of the engines used by Locomobile were two cylinders, double acting with 2-1/2 inch bores and 3-1/2 inch strokes. There was also a larger engine but this was used on only a few larger cars.

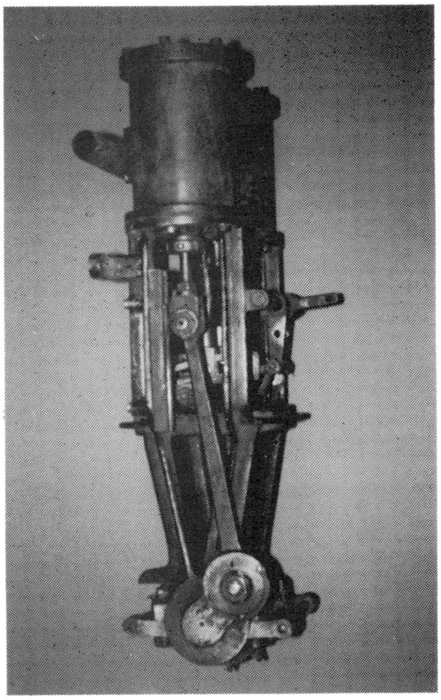

Engine Number 1 from the Mason Regulator Company. You can see the distinguishing single flat crosshead guide on each side of the engine. The frame was bronze with cast-iron cylinders.

Locomobile first used an engine purchased from the Mason Regulator Company. This I call Engine No. 1. The distinguishing feature of the Mason engine was the brass frame (the later engines made by Locomobile had forged steel frames). This engine also had flat crosshead guides on one side of the crosshead only. There was one socket for the water pump.

Side view of Engine Number 1 more clearly illustrating the single-side crosshead guide.

Early in 1900, the Company developed a first engine of their own (Engine No. 2). The cylinders were 4-7/8 inches long while the pistons were 5/8 inch thick. There were two piston rings which locked together with a pin in one ring entering a hole in the other. This engine was used up to about serial number 2100.

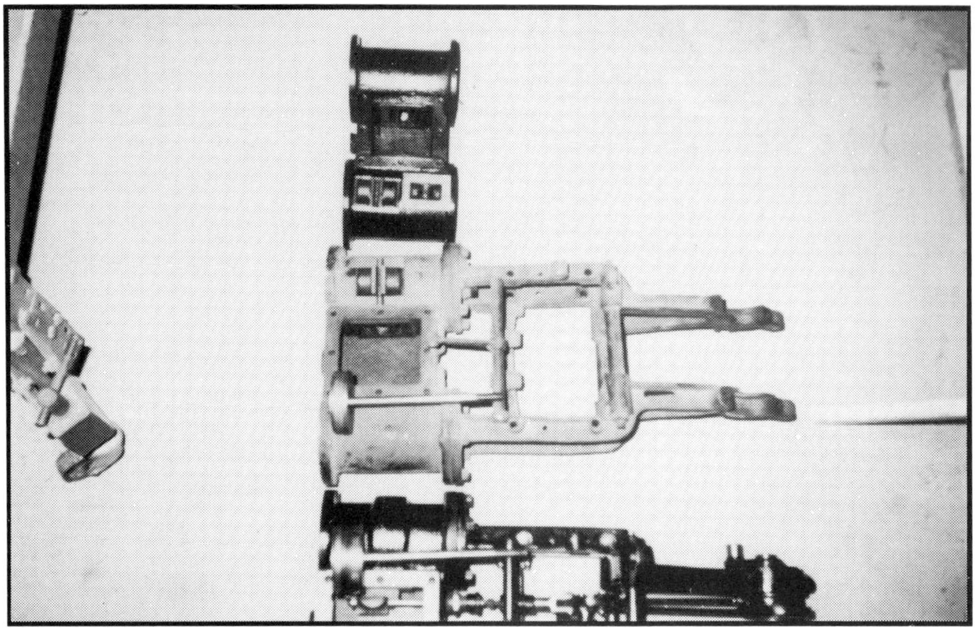

At the top of the picture is the cylinder block for Engine Number 2, the middle is Engine Number 3 while the bottom is Engine Number 5. You can see the difference in the length of the cylinders between Engine Number 2 and the others. The two pistons lying on the engine and the two valves show also the difference in the thickness.

The piston thickness was apparently a problem as the next iteration of the engine produced a cylinder 5/8 inch longer than the first with the piston thickness being the same. Outwardly, the rest of this Engine No. 3 was pretty much like the Number 2.

On the ground is Engine Number 5. The webs added at the water pump bracket and the pump pivot bracket can be seen. Engine Number 2 did not have these webs and breakage was likely at these points. The band added around the cylinders was to provide for increased operating pressure in the later cars.

Engine No. 4 was developed apparently due to stress on the mount for the water pump. Engine Nos. 2 and 3 had only a flat piece of steel with a hole in it jutting outward from the frame. Through use, this piece probably fatigued and broke off because No. 4 had the addition of a small flange which reinforced this mount.

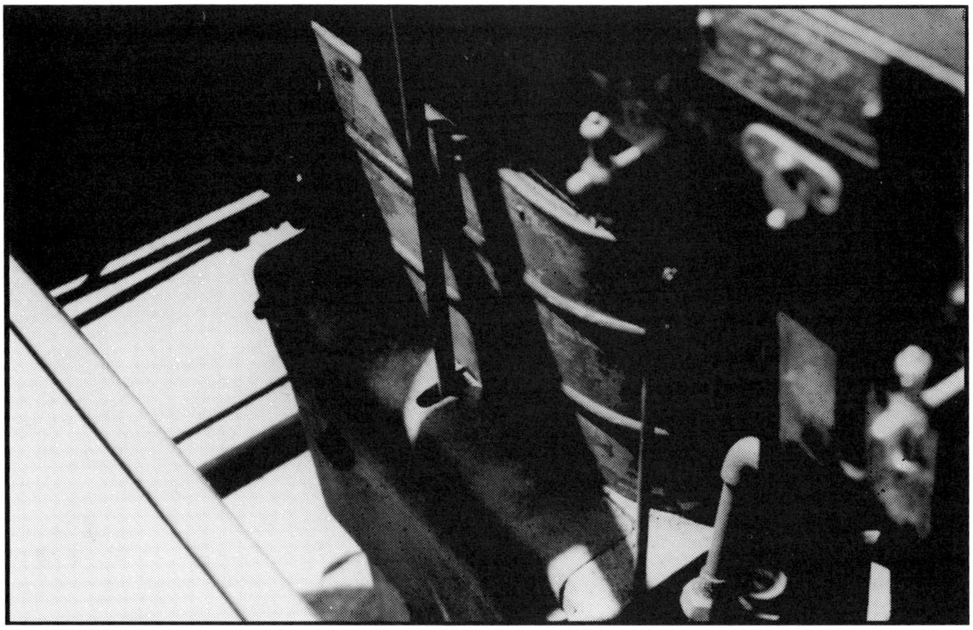

Engine Number 6 was similar to Engine 5 except that it was encased in an aluminum jacket ending some of the earlier problems with dust. The door on the front allowed access to packing glands.

Engine No. 5 was identical to No. 4 except that higher steam pressures dictated higher strength cylinders and a band encircling each cylinder was cast into the part. This engine was first used about serial number 3100 and remained in this configuration until probably the end of production.

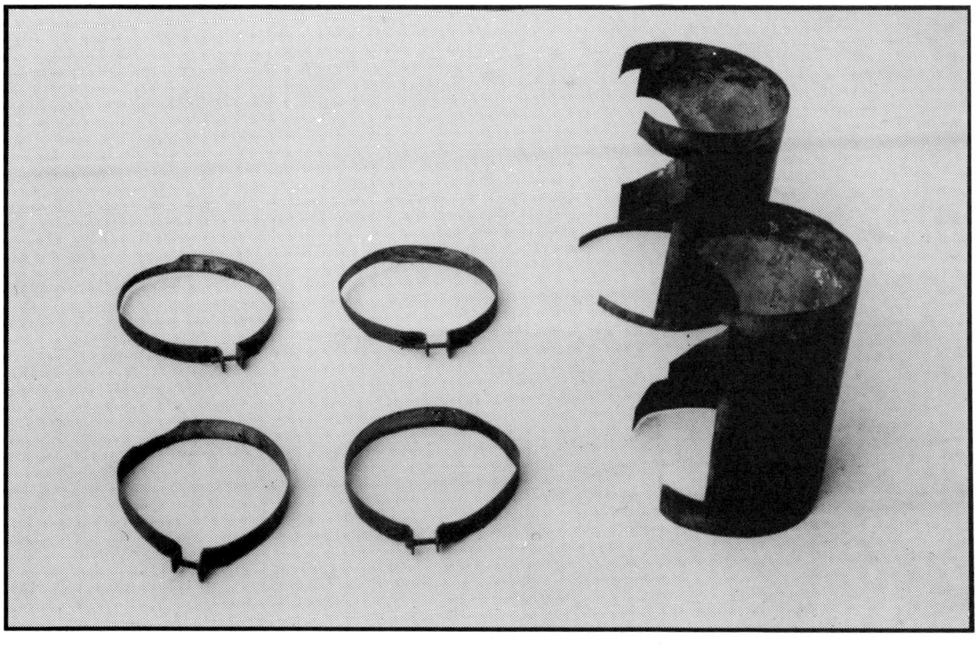

The early engine covers were just brass covers over the cylinders simply held on by brass clamps.

Engine No. 6 had larger cylinders (3 x 4) and was entirely enclosed in an aluminum case. In addition, the packing gland nuts were provided with small handles so that they could be tightened without any undue problems.

The later engine covers housed the entire cylinder block thus allowing more insulation to be installed and better efficiency. The displacement lubricator can be seen mounted on the left side of the engine block with a valve regulating oil flow. This lubricator was first used in 1901.

Until about 1902, the engine and boiler were supported by a wood framework extending from side-rail to side-rail. In 1902, a prefabricated metal frame was introduced which was bolted to the wood side-rails of the body. The engine mounts were bolted directly to this frame as were the boiler mounts.

Lubricators

The cylinder lubrication was provided by a small displacement lubricator attached to a pipe leading to the steam chest on the earliest cars. This lubricator carried enough cylinder oil to last 35-50 miles. By late 1900, the lubricator had grown larger and had a range of about 100 miles. The larger size necessitated bolting it to the engine cylinder head by means of two of the cylinder bolts. It was still of the displacement type, however.

The old displacement lubricators varied in efficiency depending upon the steam temperature, the atmosphere and other variables. This led, in 1902, to a Rochester lubricator activated by an arm attached to the crosshead bolt which force fed oil into the cylinders at a steady rate.

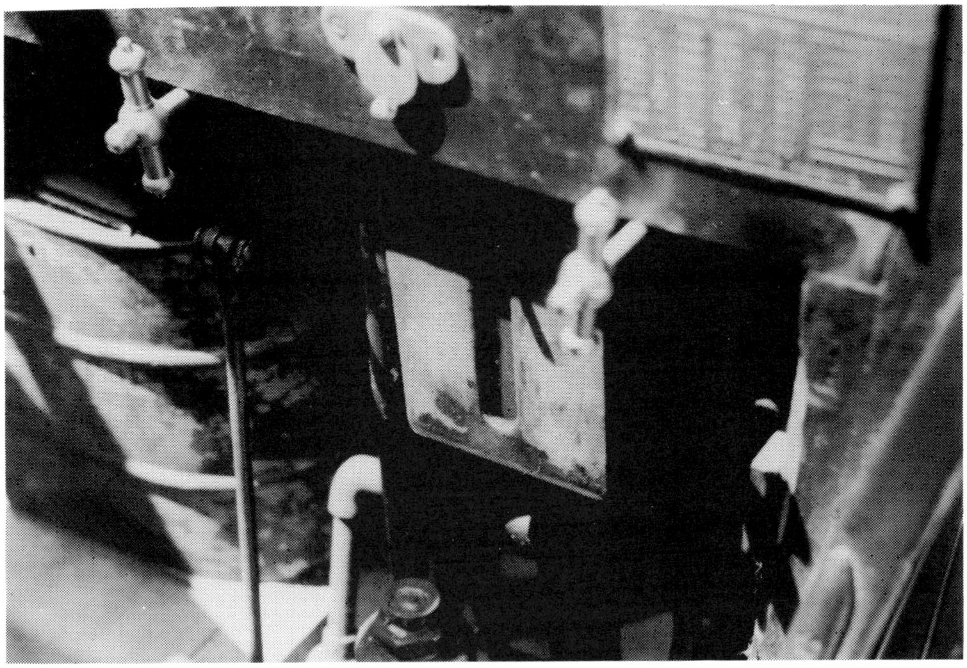

After 1901, a Rochester lubricator was mounted to the engine and driven off the crosshead guides. This forced oil into the cylinders a lot more efficiently than the early displacement types.

Engine Water Pumps

The first water pumps used were on the Mason engine and manufactured by Mason. They consisted of a round cylinder with separate inlet and outlet angle check valves screwed to the cylinder.

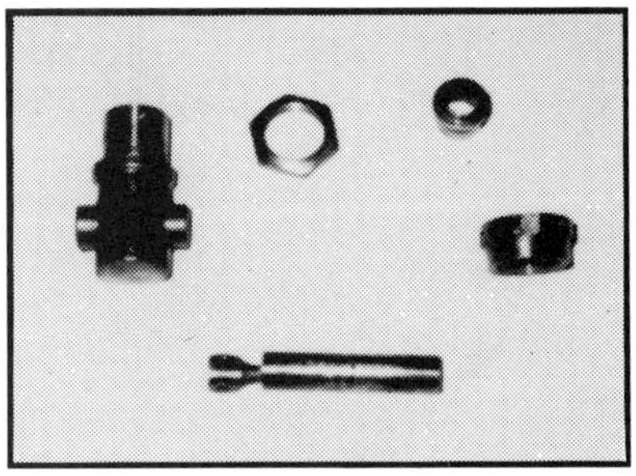

Water Pump Number 1 was simply a cast cylinder with check valves (not shown) added on the inlet and outlet sides.

When Locomobile started making its own engine, it used a similar water pump except that the pump casting was in one piece including the check valves.

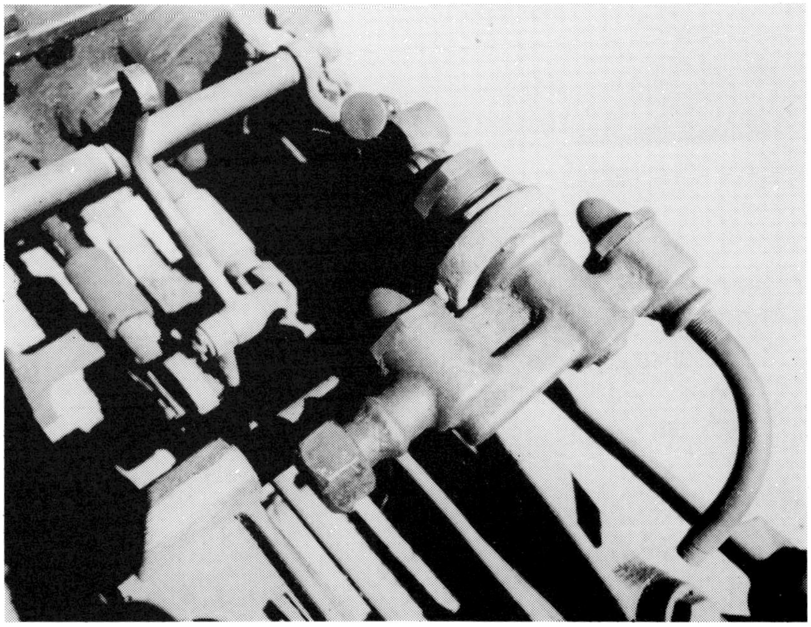

Water Pump Number 2 had check valves cast integrally into the body casting.

By mid-1900, a new water pump had been devised and had the appearance of a large and a small cylinder side by side in one casting. The smaller cylinder contained the check valves and were unique in that the stem of the outlet check nested inside the inlet check. This pump was apparently fairly successful and was used until sometime in 1901 when a new pump was developed.

Water Pump Number 3 had check valves in the cylinder on the left. The inlet and outlet valves were nested with the outlet check valve having a hole in it to serve as the guide for the inlet check valve.

This new pump looked like an tee and had the inlet water line entering from the bottom of the tee with the outlet being at right angles to the main line of the pump. The plunger, of course, entered from the top. This pump was used until the end of production, a larger version being used with the larger (3 x 4) engine.

The pump plunger size started at 1/2 inch and increased to 9/16 inch and finally to 5/8 inch on the last pump variation.

Water Pump Number 4 had check valves integral with the pump body. The inlet was on the bottom with a ball check while the outlet was on the piece coming out the side. A union held a spring-loaded ball to serve as the check valve.

From the left: a style 5 Locosurrey and two style 02 or 3 Stanhope steam carriages. Taken in front of a dealership apparently in the Far East!

Original Factory Photo.

Burners

Locomobile's burners did not change much over the years. A burner plate consisting of two pieces of sheet metal about an inch apart was standard equipment. Numerous half-inch holes were located in each plate with the plates connected by short lengths of tubing, similar to boiler tubes. Around each tube on the top plate several small holes were drilled. In operation, the fuel/air mixture entered between the plates and exited through the small holes. As it exited, the mixture sucked air through the tubes and helped create a hotter fire.

Top view of the early burner showing air holes surrounded by tiny gas jets holes.

Bottom view of the same burner showing the air holes.

The early steam automatic with the taller valve to the right controlling the main burner flame and the other controlling the lighting torch.

The lighting torch. The angle valve handle at the left served only to screw the hot torch into the steam automatic. The valve on the automatic controlled the fuel flow. The jet of the torch is at the end of the pointed section at the bottom.

A detachable torch was used to start the Locomobile. This was a bent piece of tubing with a small hole for a fuel jet in one end and a screw fitting at the other. This torch was taken off the car and heated in a flame, over the kitchen stove perhaps. While still hot, the bent part of the torch was inserted into the burner through a slot provided for it and the jet hole pointed into the mixing tube. The screw fitting engaged a hole in the main fuel line and was tightened. When the fuel valve was opened, the torch, hopefully, was still hot enough to vaporize the gasoline, and the burner could be lighted. The torch was used until the vaporizer tube in the boiler got hot enough to vaporize the gas and, when the corresponding valve was opened, the burner would light using the main vaporizer. The lighting torch could then be shut off, withdrawn from the burner and put in the tool box for the next time.

No pilot light was used and the steam automatic would merely shut the burner down to a low flame rather than shutting it off entirely.

By around 1902, a small pilot was introduced into the burner. This consisted of a cup about two inches in diameter with a couple of loops of tubing in it. The cup extended below the bottom of the burner so that gasoline or alcohol could be put into the cup, being absorbed by cotton waste, and lit without opening the burner. When the pilot was hot enough, the main burner valve could be opened and the burner started. It's interesting to note that the starting torch was retained for emergencies, thus indicating the faith the designers had in the pilot.

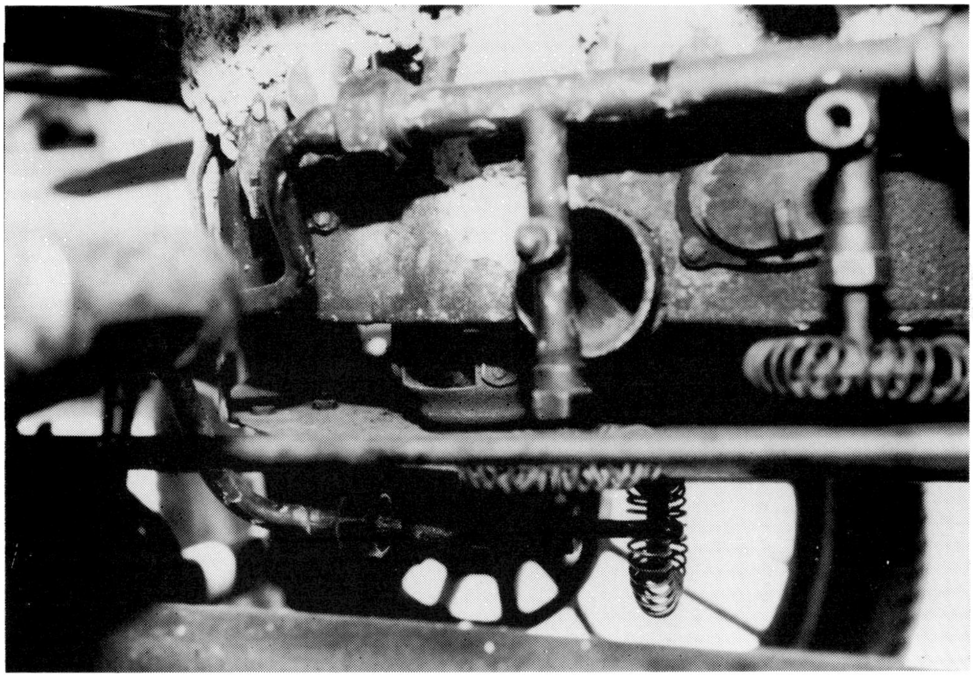

The late-model burners were essentially the same but with a pilot which can be seen protruding below the bottom of the burner. Gasoline was allowed to run in here and was lit to preheat the pilot. Below the pilot can be seen the flat metal sheet used to protect the bottom of the burner from road dust and mud. The small oval door between the valve and the venturi tube is for the insertion of the lighting torch which could still be used in an emergency.

The rest of the burner remained the same except the later cars (1902 and up) had a baffle over the bottom of the burner so that mud and water could not be thrown up into the burner pan.

The early fuel shut off was located on the left side of the car under the body which made it necessary for the driver to get out of the car to shut down the fuel. In 1900, this was changed and the valve was relocated below the reverse/throttle lever quadrant allowing the driver to shutoff the fuel without leaving the seat.

Like many other vehicles, however, rules can be broken. There is an original 1902 Stanhope B in existence with the fuel shutoff valve still on the burner.

Throttles

The Style 1 car was first equipped with a poppet valve throttle connected to the boiler. This worked well except that, if the poppet valve got any wear, steam would leak and cause a stopped vehicle to eventually move off under its own power. This early throttle can be identified by a 1/8 inch pipe thread outlet which was used to connect the safety valve for the boiler. A threaded hole was provided to connect a stay brace from the throttle to the front seat crosspiece.

The early throttle was braced from the body with a rod flattened on one end with a hole in it. This rod was screwed to the throttle body by the screw on the left. The safety valve was attached at the threaded hole at the bottom right. The threaded portion at the top of the body held the packing gland from which the throttle valve protruded.

The boiler safety valve on the early cars exhausted into the air. Later models had a line leading to the water tank to help muffle the noise. The lever on the left of the valve was for testing the safety valve.

The possibility of a leaking throttle was quickly corrected in 1900 when an auxiliary throttle was provided between the boiler and the regular throttle. This auxiliary throttle was simple a globe valve with a quick-acting stem (1-1/2 turns open to close). When leaving the car, the auxiliary throttle was closed thus preventing steam from leaking through the poppet valve. The valve handle was made removable by means of a thumb screw and could be taken with the driver as sort of a primitive key.

By early 1900, a new throttle was devised which was essentially the same as the early one except that the threaded hole for the stay brace was replaced with a clear hole located in a bracket cast into the body casting. The pipe thread for the safety valve was removed as the safety valve had been relocated to a line from the boiler following the addition of the auxiliary throttle.

The later throttles used a brace rod threaded on both ends which screwed into the body by the bracket on the bottom. The valve rod now had a swivel job to attach to the throttle lever. Note that the safety valve no longer screws into the throttle casting.

The later throttle (on the left) and the auxiliary throttle valve which positively shut off the steam flow. This was simple a globe-type valve between the boiler and the throttle.

This flexible swing joint was used to connect the throttle to the engine. The elbow on the left screwed into the engine and the joint screwed into it and to the throttle. Packing was used for sealing the joint.

Rear Axle Housings

The earliest rear axle housings were made, as were the later ones, with steel tubing but with two flat U-shaped pieces attached in front of and behind the chain sprocket. The pieces held the right and left halves of the housing together. After some usage, the flat pieces tended to bend, opening the U and allowing the differential gears on the ends of the axles to disengage from the spider gears. When this happened, of course, the car came to a stop.

The early rear end with the flat U-shaped plate joining the two rear axle housing halves. The differential gear was exposed and subject to road dust, etc. Since there was little lateral bracing, the two housing halves tended to spread apart, disengaging the differential gears. The brake shown is an accessory.

The first attempt to fix this problem consisted of applying two braces extending from the innermost end of the axle tubes to the bow-shaped piece of the axle housing above. These braces were merely clamped on the tubing and a right-and-left thread was used to adjust the tension.

The first attempt to solve the axle spreading problem consisted of using round tubing to connect the two housing halves plus two clamp-on braces to help hold everything together.

The clamp braces fit around the axle housings and the bowed section of the axle housing. A right/left thread arrangement allowed tightening to the necessary pressure.

This apparently did not prove to be entirely satisfactory and the flat U pieces were replaced with U-shaped pieces made of tubing. The braces were retained, however, until late 1900/early 1901 when the axle was redesigned to include the braces permanently attached to the axle housing. It seems that the flat pieces were replaced with the tubing in early 1900. These are quite common on Style 2 cars but at least one Style 1 car has survived with the round tube pieces on the rear axle housing.

The axles themselves originally had threaded ends with matching rear-wheel hubs which screwed on the axles. They were right-hand and left-hand threaded respectively so that each stop of the car tightened the wheels on the axles, making them very hard to remove. Sometime in 1900, these axles were replaced with ones having square ends which fit into hubs having square holes to receive them.

The differential gearing on the early cars (until 1901) consisted of spider gears on a carrier which also held the chain sprocket and differential gears on the ends of the axles mating with these. This is all very similar to modern differentials.

In 1901, a new differential housing was introduced which consisted of a spur gear differential system, all enclosed in a cast iron housing. The ends of the axles were now equipped with spur gears mating with other gears inside the housing. It is interesting to note that the housing had no way of getting grease into it. Apparently, you filled it with grease when it was initially assembled and no loss was anticipated.

In 1901, a new spur gear differential was introduced which encased all the working parts in a cast iron housing. The brake band width was made larger at the same time. By now, the rear axle housing had permanent braces to replace the temporary clamp-on braces used previously.

The interior of the spur gear differential showing the axle gears with the smaller gears below.

In about 1902, another type of differential was introduced which appeared to return to the earlier type but all was enclosed inside a cast housing. I have never seen the inside of one of these differentials but they were very compact.

The most advanced differential involved two brake drums, one on each side of the driven gear. The differential gears were encased in a housing. The two brakes were double-acting.

Drive Chains

At first, the cars used a 1 inch x 5/16 inch block chain (example below) with a 2.5:1 reduction between the engine and rear axle (a 3:1 ratio was used on the four passenger surreys). By mid-1900, the roller chain was in general use on the cars. It measured 3/4 inch x 5/16 inch and is no longer available as a standard size. There is evidence, however, that block chain was used on the surreys into 1901.

The chain tightener consisted of a Y-shaped piece fitting on the front part of the rear axle with a tube which connected to a smaller Y-shaped piece on the rear of the engine. These were connected by a threaded piece which allowed the length of the tightener to be adjusted to the chain and slack in the chain thus be removed. The chain tightener simply held the rear axle and the engine sprockets a fixed distance apart independent of the body and chassis springing.

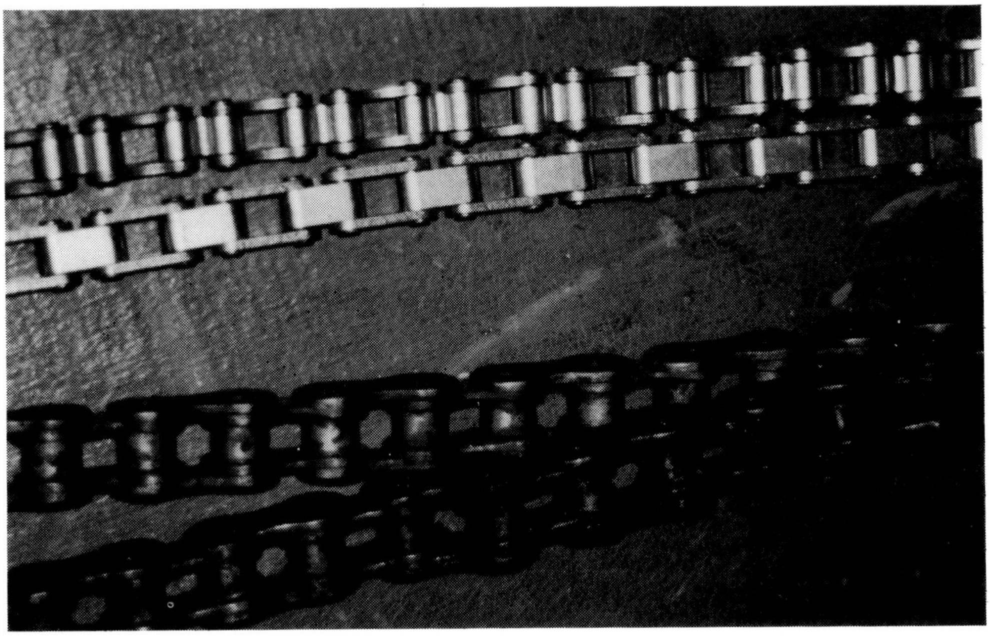

The early 1 x 5/16-inch block chain is on the top while the 3/4 x 5/16-inch roller chain is on the bottom. The block chain had much more friction in it and broke much more easily than the later roller chain.

The front half of the chain tightener was threaded and the rod connecting the front and rear parts threaded with a hex fitting so the two parts could be spread apart to tighten the chain.

The early chain tighteners on the 1899 and 1900 cars were fairly light weight with heavier castings appearing with the 1901 vehicles. The basic design remained the same throughout production.

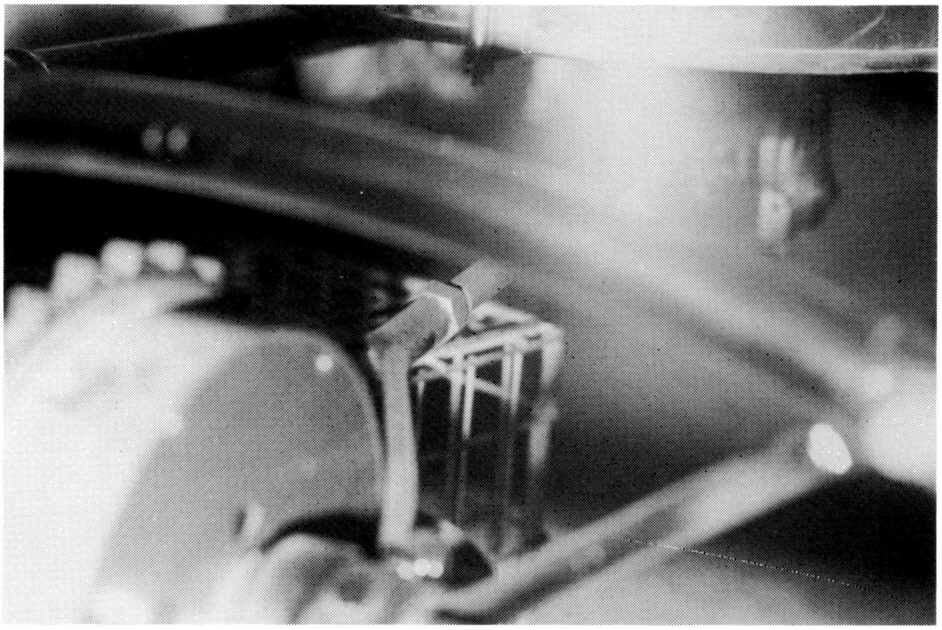

The rear portion of the chain tightener straddled the differential housing and fit into lugs bolted on the axle housing.

The front of the tightener fit into slots in the engine.

Brakes

The earliest brake on the Locomobile was a single-acting band operating on a drum immediately adjacent to the chain sprocket on the rear axle. Single-acting meant that it would only stop going in the forward direction and would not work when backing. The position next to the chain sprocket also meant that the brake surface was copiously oiled, further decreasing the stopping capability.

The single-acting brake was simply a band anchored at one end and wrapped around the brake drum. The brake rod pulled on the end of the band and applied the brake. Going forward, the band had "wrapping" action and helped to stop the car but, in reverse, there was no wrapping and, consequently, no brakes.

The double-acting brake introduced a lever connecting the ends of the brake band which allowed wrapping of the band no matter which way the car was moving. The pivot of the lever was then the anchor point of the band.

In 1901, the brake was redesigned so that it now stopped going in reverse also. This feature was highly touted in the 1901 catalog.

By 1902, most cars were equipped with two types of brakes. One type was still adjacent to the chain sprocket on the rear axle but now there were two drums, one on each side of the sprocket. These brakes were double-acting. (See photo on page 20.) There was also a set on the wheel hubs. Each set was activated by a separate pedal.

Due to the single-acting brakes, many companies provided after-market brakes which added the double-acting feature. The Dayton brake was a particularly good design. The driver could actually lock the rear wheels quite easily.

In the early years, the cars did not come equipped with any kind of a parking brake and the after-market people did well here also. The Victor brake was the most common and came in several varieties with small and large oval pedals as well as rectangular ones. By 1902, the Locomobile Company had bought the Overman Company, makers of the Victor brake and Victor steam pumps as well, and this was provided as standard thereafter.

Control Levers

The Style 1 control levers for the throttle and reverse were fairly simple. The reverse lever was a cast piece with a spring-loaded ball detent acting on a curved quadrant. The detent was supposed to lock the reverse lever into one of several indentations in the flat side of the quadrant. The throttle lever was simply a piece of flat stock extending up through to the top of the seat rail.

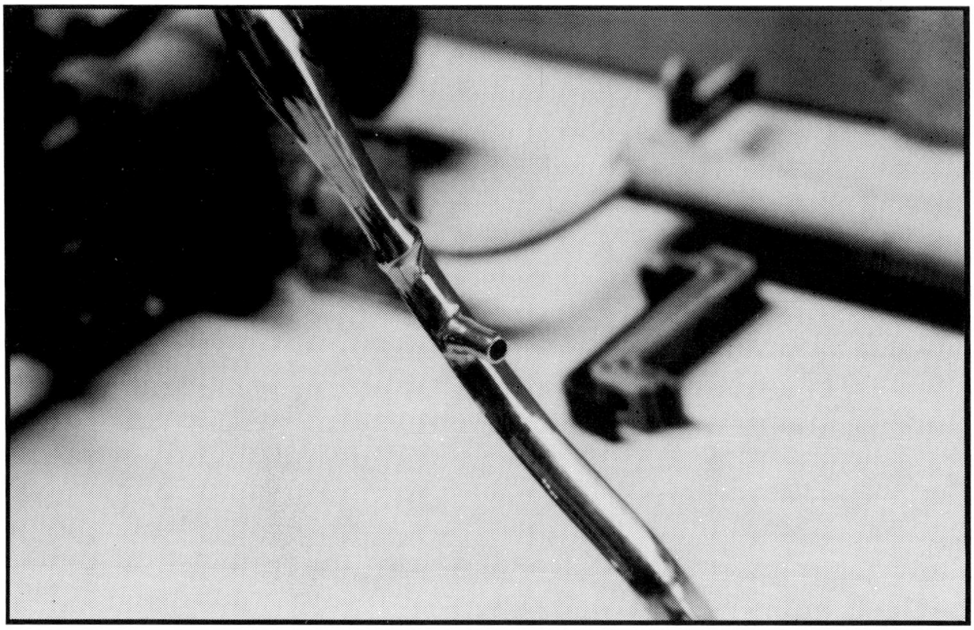

A ball-detent on the early reverse lever helped keep it in the required position for "hooking up" or reversing. The handle was cast integrally with the lever while a spring and ball fit in the protruding boss.

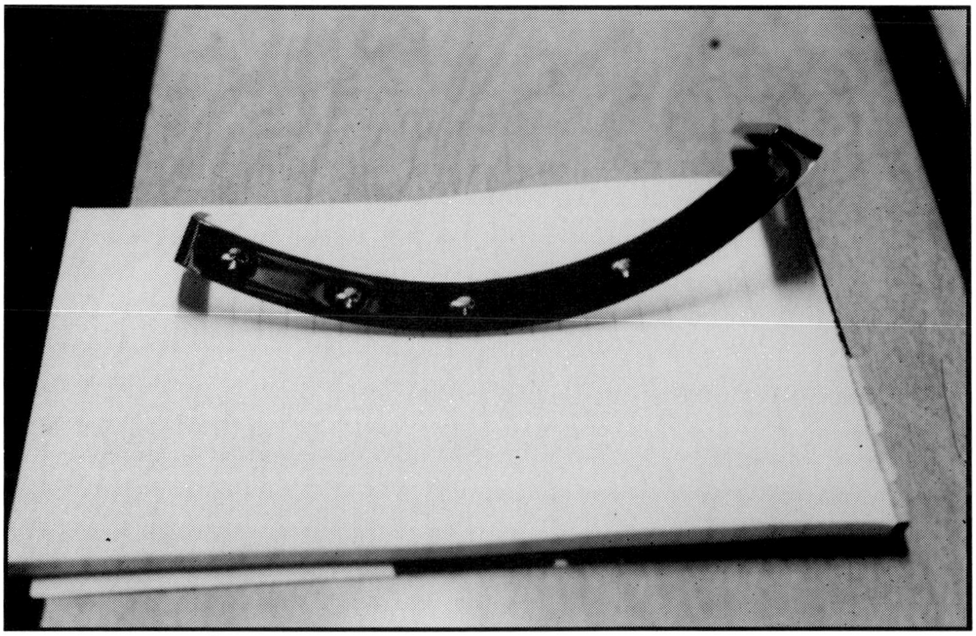

Indentations in the corresponding early reverse quadrant allowed the ball on the lever to lock into position.

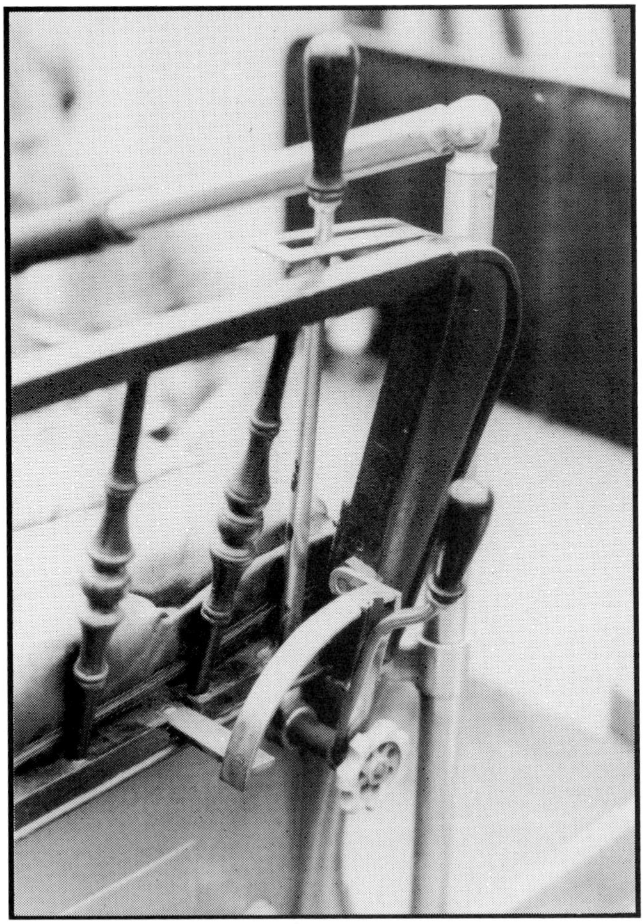

By 1900, the throttle operated in a channel which contained a notch to help lock it in the closed position. The handles were now wood and screwed onto the levers.

By mid-1900, both levers were cast pieces with separate wood handles that screwed onto the metal castings. These handles were basically round. (The Mobile company did the same thing but their handles are distinctively flatter in shape, with inlaid nickel bands.) The reverse lever quadrant was now turned so that an angle-shaped piece of spring steel rode on the edge of the quadrant instead of the flat side. This spring would engage notches on the quadrant to hold the lever in the various positions.

A spring piece on the reverse lever now fit into notches in the reverse quadrant. There were two or three notches for hooking up the engine. The water bypass lever protrudes from the center of the reverse lever while the detachable auxiliary throttle handle can be seen below.

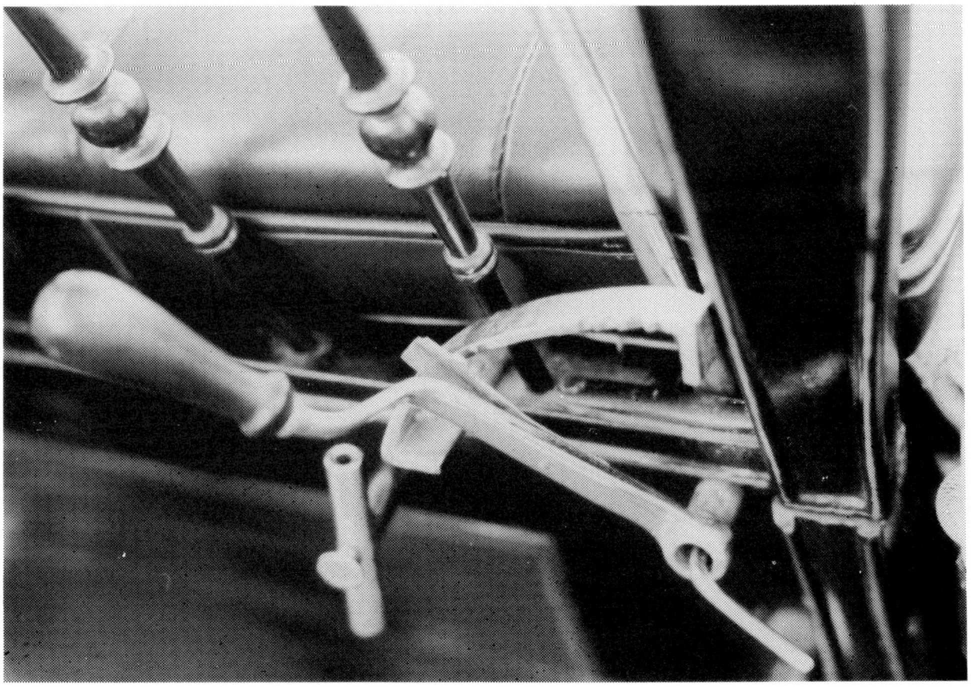

On the later cars after about 1902, the reverse lever quadrant was provided with a ratchet surface and the lever itself had a dog which fit the ratchet and was released by a handle on the lever itself, much like a miniature parking brake handle.

By 1902, certain models of the cars had this ratchet which gave a very positive hookup and reverse position. Finer adjustment of the hookup could be easily maintained. The slightly curved piece to the left of the handle is an accessory gradimeter.

Front Axle

Front axles did not change too much over the years. They kept the same basic design but became more sturdy as the cars got larger.

The first cars had the steering tiller located in the center of the car and the tiller rod connected to the tie rods. This had the disadvantage that, due to the front spring, the tiller moved up and down in relation to the car body. The tie rods on these early cars were 1/2 inch diameter with ends that were merely slotted to fit over the steering arms.

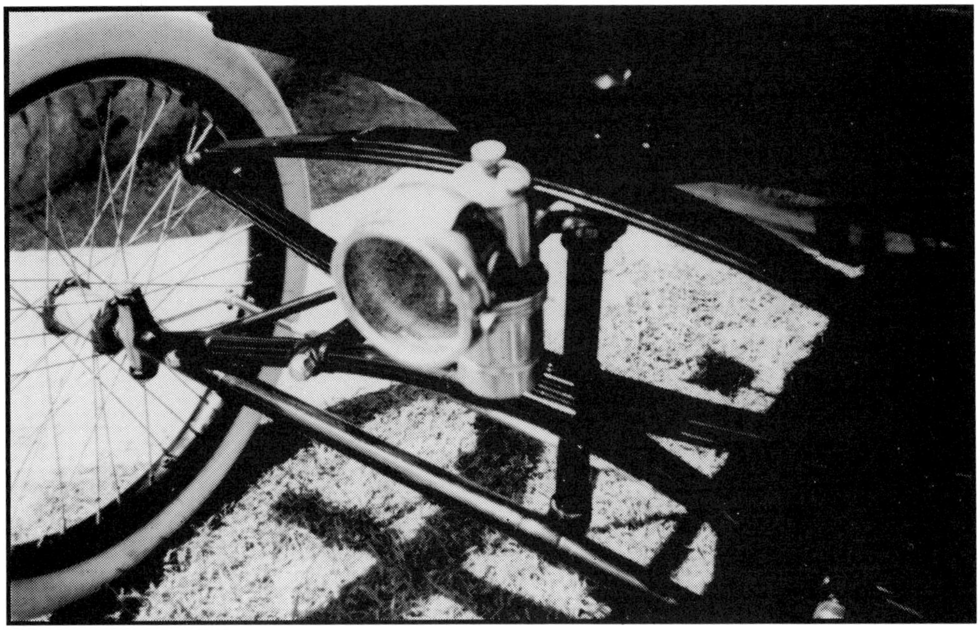

Early tie rods were 1/2 inch in diameter with very small ends which attached to the steering arms with a screw.

In early 1900, the tiller was moved to the right side and connected with the tie rods through a drag link which passed from the tiller column under the car to a crank located on top of the front axle. By this time, the tie rods themselves were heavier (5/8 inch) and the ends were shaped more like a proper clevis with clevis pins holding them to the steering arms.

The tiller pretty much continued throughout production although a few models had a steering wheel located on the right side of the car.

Later tie rods were 5/8-inch in diameter with a heavier clevis attaching to the steering arms with a clevis pin.

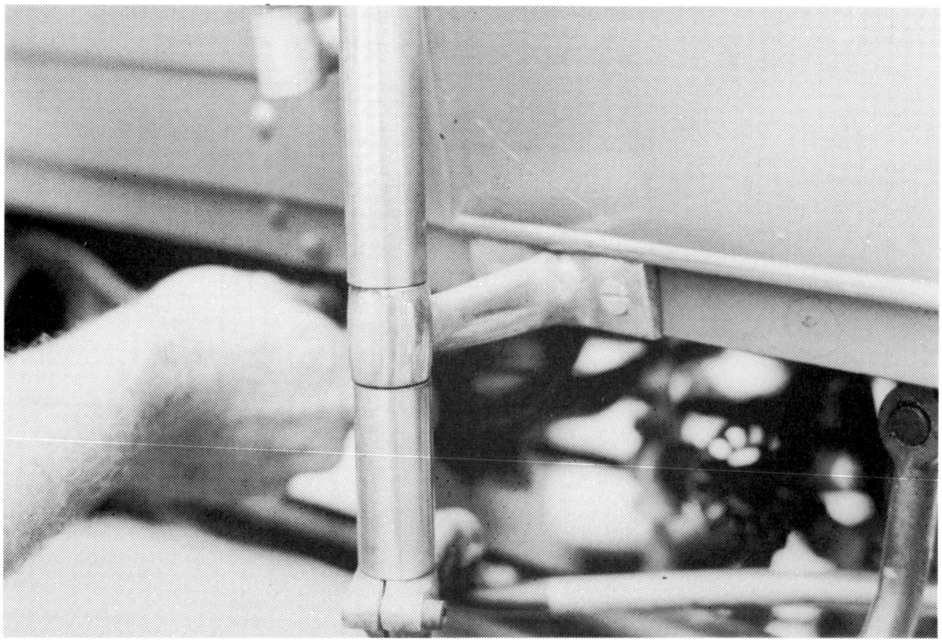

When the tiller was moved to the side of the car, a drag link connected the bottom crank to the front axle. The tiller necked down to fit in the bracket and had a sleeve covering the smaller diameter below for cosmetic purposes.

The upper end of the tiller bolted to the body with this bracket.

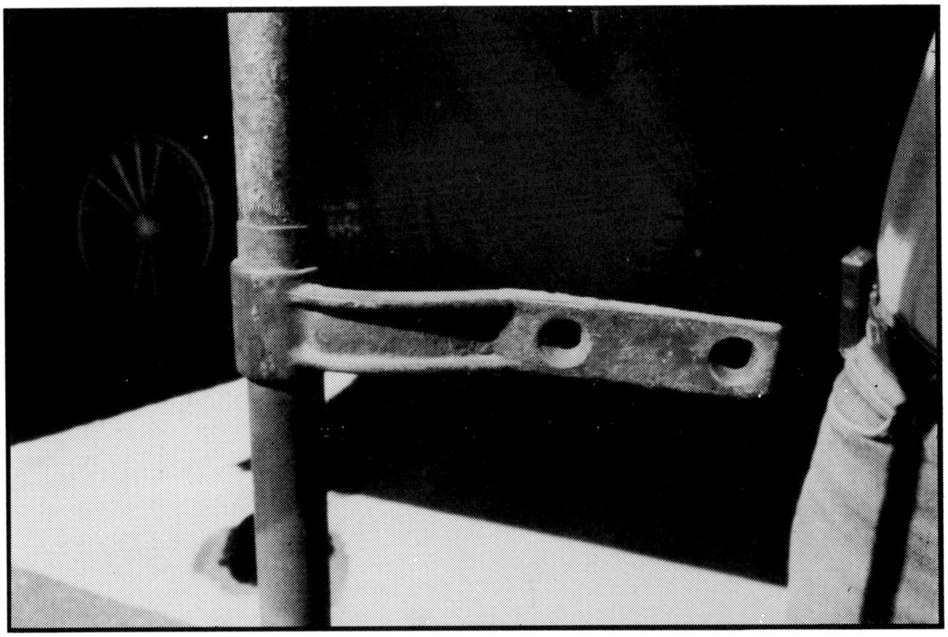

Some of the earlier models used a two-hole bracket for the upper tiller mount.

Tire Tread

The tread on the Style 1 cars was about 51 inches. The Style 2 cars had 54-inch treads while some of the later, heavier cars such as the Touring Models A and B had 56-inch tread.

Chassis

The chassis of all the Locomobiles was basically the same. It consisted of tubular steel parts furnace brazed with steel forgings to form the front and rear axles and the two reach rods. The reach rods were designed so that the two axles could pivot a small distance with respect to each other and thus allow flexure. The size and length of these pieces varied, the car model with longer reach rods accommodating larger vehicles.

The springs on the early two passenger models were made of 1-3/8 inch wide steel and the cars were provided with two full-elliptical springs on the rear axle and one full-elliptical transverse spring on the front axle. The surreys had 1-1/4 inch wide springs with two on each axle.

By 1902, the larger cars such as the Stanhope B, Dos-a-Dos, Touring Models A and B and the Long Wheelbase runabouts all had four full-elliptical springs. (See photo on front cover.)

Bodies

Locomobile made several different body types over the years (see below and Appendix). The earliest bodies were of the spindle-seat stanhope type. The small deck behind the seat was covered with a leather cover and there was no smokestack. When standing, the leather cover tended to get very hot with holes eventually burned through it. By mid-1900, a metal smokestack was provided which gave vent to the burner when the car was standing. The leather cover was replaced by a metal one in 1901.

Body widths varied depending on the model and gradually grew sturdier as technology progressed and cars got larger. The Style 1 bodies were 25-1/2 inches wide while the Style 2 cars were 28-1/2 inches wide. Upholstery was usually leather but there are some documented cases of whipcord being used on early Style 1 and 2 cars. Perhaps others had it, too.

The Style 2 was manufactured from basically the beginning of production through the end. This particular car was made in late 1900.

The Touring Model B was probably the most advanced of the Locomobiles and sometimes even featured a steering wheel, larger boiler and engine with wood artillery wheels.

The early cars until perhaps the 2nd quarter of 1900 had a wood dash which was varnished with black paint. After that time, the standard dash was a metal frame covered with patent leather. A handhold was provided on either side of the dash for assistance in getting into the car.

A leather flap hung down from the front of the front seat covering the engine. Behind this flap was another leather flap bearing pockets for tools, oil cans and an air pump. The Locomobile people thought so much of this tool flap that they applied for and were issued a United States patent on it.

Since storage space was at a premium, a tool flap fitting between the engine and flap covering the engine below the seat was added. This tool flap could hold a tire pump, extra gauge glasses, wrenches, screwdrivers and oil cans. Even a space for the instruction manual. It was attached to the body by clips similar to those used on galoshes.

Colors varied. Mr. Figge's article has the best description of the standard varieties but others were known also. Perhaps Locomobile would paint it any color you liked if you paid them for it. From Figge's article, the Style 1 and 2 cars were:

1. Black chassis with black body, no pinstriping.

2. Yellow or black chassis. Black body with yellow side panels. The rear panel may be black or yellow. Black striping on yellow chassis, yellow striping on black body areas and black chassis.

3. Red or black chassis. Black body with red side panels. The rear panel may be black or red. Black striping on red chassis, red striping on black body areas and black chassis.

4. Brewster green or black chassis. Black body with brewster green side panels. The rear panel may be black or brewster green. Gold striping on chassis and black areas of the body.

Some cars had the inset area comprised of the side of the front seat, extending down and then forward to the side panel of the floor being painted the same color as the body panels. Another variation would be to paint the wood frame beneath the body, which is inset, to match the body panel color. Both of the above-mentioned variations appear to be less common than combinations 2, 3 and 4 above. Combination 1 is the most common for first year production.

Side lamps were standard on the car. The type varied considerably from kerosene lamps to self-generating carbide lamps. A small gauge lamp was provided to play on the sight glass so that it could be read by night. Several different suppliers were used and dealers and owners regularly found other automobile and bicycle lamps to replace the factory-supplied ones.

Tops

Tops varied from year to year and car model to car model (see list below for details). In general, the body was provided with irons similar to a buggy of the same period and a buggy-type top was attached to these (occasionally, a victoria top was used). The tops resulted in so much stress on the car that usually only the solid panel seat vehicles carried a top. The spindle seats just did not have enough shear strength to withstand a good wind without the top blowing off and taking the seat with it. In early 1900, however, there was a model that had a spindle seat with a top. It was very short-lived, no doubt for the reasons given above.

Side curtains were unknown but a storm apron attaching to the lower front of the body and to the back of the seat acted like a bib to keep rain off the passenger's clothing. Flaps hung down on the sides to prevent water from entering from this quarter. As long as the driver had a good hat on, he didn't get too soaked, unless he was driving a spindle-seat car which had a built-in draft around the rear.

On most models, the storm apron folded up and was stored under the floorboards. On the Touring Model A and B, the apron was on a roller similar to a windowshade and would roll out of sight under the front of the floorboards.

Accessories (siphons, steam pumps, etc.)

The first Locomobiles had very little besides the burner and boiler. Air and steam pressure gauges were standard as was a steam automatic to control the burner and a sight glass to display the water level in the boiler. A mirror mounted on top of the steam gauge allowed one to see the level in the sight gauge. There was no hand water pump, only the pump on the engine.

By 1900, a hand water pump was standard and the sight glass mirror was mounted at the bottom of the dash. The check valves on the water glass lines were now provided with valves to reset the checks. The purpose of these checks was to protect the driver against breakage of the sight glass. If steam rushed through the lines leading to a broken sight glass, these check valves would seat and stop this flow, thus allowing the driver to replace the glass. The early cars had no way to reset these valves without first blowing down the boiler. These manual valves allowed resetting without blowing down.

The early-style water pump had a relatively small (5/8 inch) plumber and was equipped with separate check valves.

The later hand water pump had a 1 inch diameter piston with folding handle and integral check valves in the pump body.

Most cars had a standard sight glass for boiler water level. The glass was either 1/2-inch or 5/8-inch in diameter.

In 1902, after the Locomobile Company bought out the Overman Company, manufacturer of the Victor Steam car, Victor steam-operated air and water pumps started to appear on the larger cars and, by 1903, were on all models including the basic Style 2. A steam syphon was now standard also and relieved some of the drudgery of hauling water in a collapsible bucket.

Victor or Overman pumps were installed on better models by 1902. The pump in the foreground was the air pump (with the larger cylinder) while the one in the rear was for boiler water (smaller cylinder). Both pumps were steam operated and based on the Westinghouse design then standard on railroad locomotives.

A steam siphon (upper center of picture) was standard on most cars by 1903.

Also standard by 1902 were lubricators oiling the moving engine parts such as the cross-head guides and main bearings. The first method of doing this consisted of a brass reservoir strapped to the engine cylinders. Leading down from this reservoir were several brass pipes leading to various spots on the engine. These pipes were filled with a wicking material which normally did not allow the oil to flow. When the car was in use, however, the wicks heated up and the oil flowed relatively freely. Later in 1902, a lever-operated pump worked by the engine was added to accomplish the same purpose.

Again, by 1902, the old round sight glass was replaced by the Klinger gauge, a reflecting water glass made on the Lunkenheimer patent which made reading the water level much easier than before. A dial-type water gauge was also provided on the supply water tank.

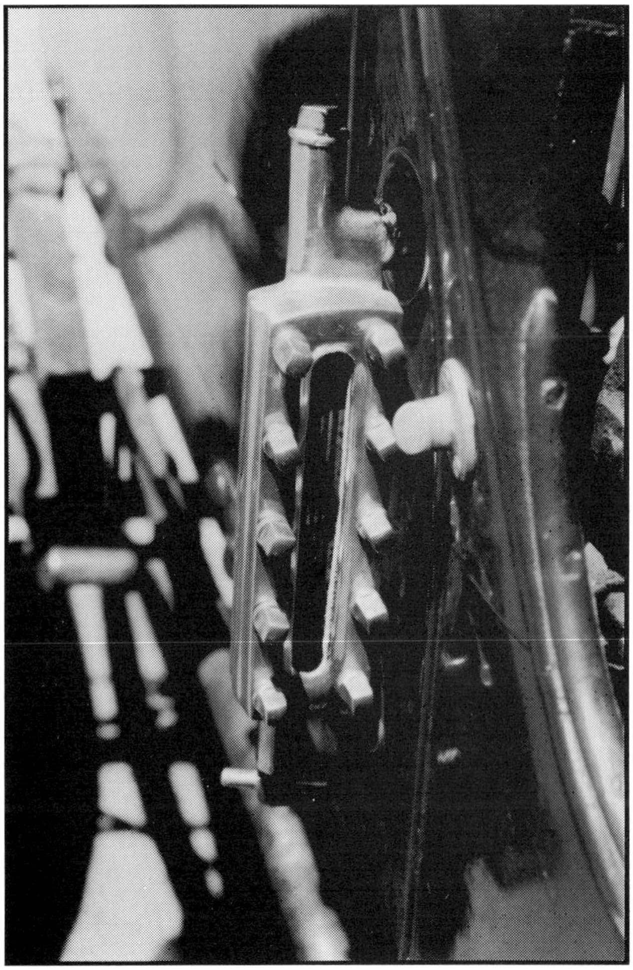

By 1902, the Klinger reflecting boiler-level gauge was standard on the higher priced cars and standard on all by 1903.

From the early Style 2 models, a Veeder cyclometer measuring distance only was provided either on the front steering arms or on the rear axle housing. Both were activated by a pin attached to a spoke of the wire wheel or a pin screwed into the wood wheel spoke.

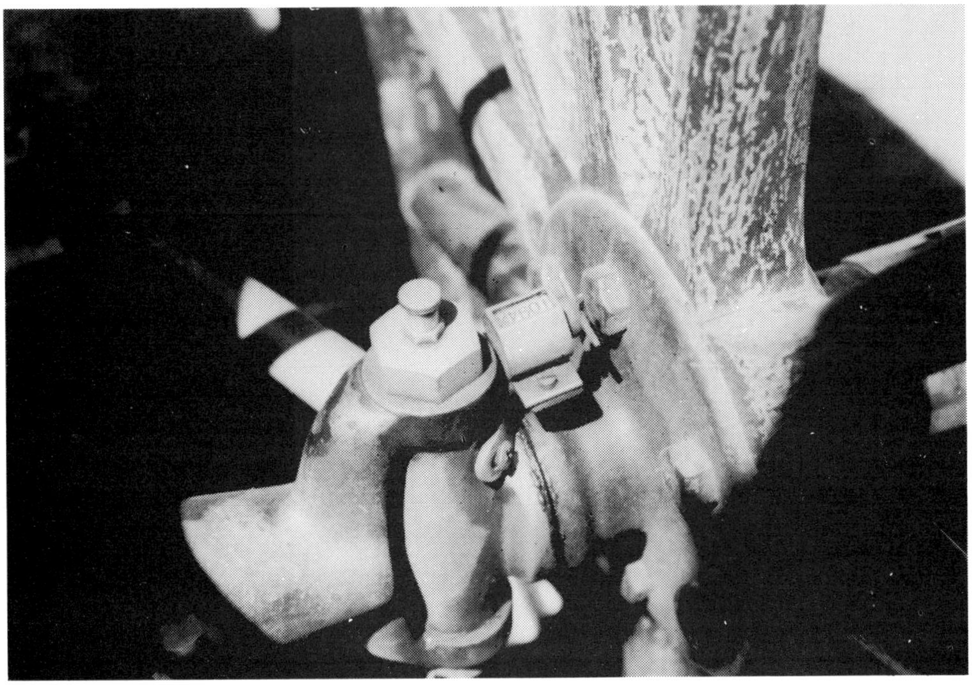

The Veeder odometer was standard from early 1900 and fit either on the rear or front wheels. This model is mounted on the front axle with the star wheel being activated by a pin screwed into the hub flange.

Through part of 1900, the exhaust left the engine and passed through a muffler before exiting through a hole in the water tank. The end of the exhaust pipe had a nozzle attached to it which produced a vacuum, thus keeping a good draft on the burner. By mid- to late-1900, this muffler was provided with a coil of brass pipe for the feedwater to pass through, thus preheating it and giving greater efficiency to the powerplant.

To muffle the exhaust was the first purpose of the muffler. By 1901, a coil of 1/4 inch brass pipe was installed inside the muffler to provide a feedwater heater. Exhaust steam from the engine entered the muffler through the fitting on the side and exhausted through the top fitting. The water entered one fitting on the bottom (left) of the muffler and exited through the other en route to the boiler.

The interior of the muffler with the coil of tubing. This gave a better efficiency to the power plant.

Water tanks before 1901 had flat tops and set a couple of inches below the top of the body to allow room for the flue from the boiler top.

An extension for the water filler cap was needed to bring it up to the level of the rear deck cover.

In 1901, the sides of the water tank were raised to allow greater water capacity and still allow room for the flue.

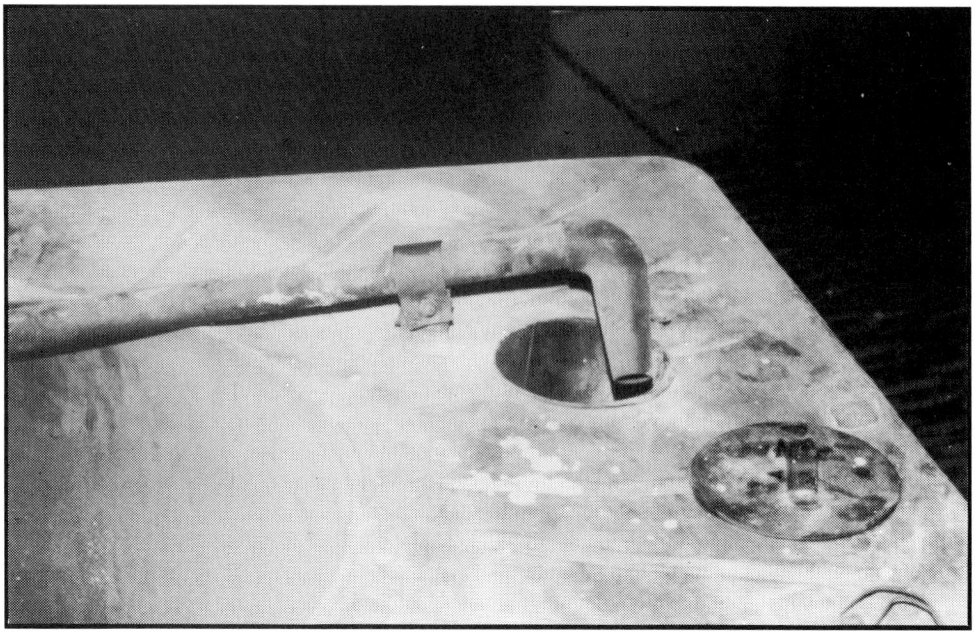

The engine exhaust went through this nozzle down through a tapered hole in the water tank. The combination of tapers produced a venturi which gave draft to the burner. The hot steam also tended to pre-warm the water.

A strainer fit into the neck of the water tank filler cap to prevent leaves, dirt and other objects from getting into the tank and later clogging a check valve.

APPENDIX:

"Locomobile" Catalogue Offerings by Year

From the left: style 05 Surrey, style 03 Victoria Top and another style 05 Surrey, taken in front of a Locomobile dealership. Original Factory Photo.

1899

Model	Body Style	Notes
1	Stanhope w/spindle seat	First model produced starting late November or early December
2	Stanhope w/spindle seat	Started in December with wide body and wheel tread. 5 gallon gas tank and 21 gallon water tank.

1900

Model	Body Style	Notes
2	Stanhope as above	Early models with center tiller. Later ones with side tiller.

"Locomobile"

Style No. 2

Standard

Price . . . $750

f.o.b. Bridgeport, Conn.

Data

Seating capacity	Two passengers
Wheels	28 inches in diameter; steel spokes
Tires	2½-inch single-tube pneumatics
Tread	4 feet 6 inches
Weight, empty	640 pounds
Weight, tanks filled	850 pounds
Capacity of gasoline tank	5 gallons
Capacity of water tank	21 gallons
Extreme length	7 feet 4 inches
Extreme width	4 feet 10¾ inches
Extreme height	5 feet 2¼ inches
Seat	Spindle back

Equipment

Rubber bucket, side lamps, gong, cyclometer, full set of tools.

NOTE—The illustrations of the different styles show the engine covered. Our experience and that of our customers has shown that this cover is unnecessary and is in the way. However, an engine cover will be supplied on request. This applies to all styles.

1900

Model	Body Style	Notes
3	Stanhope w/spindle seat. Also came with panel seat sides as same model number.	Provided with victoria or buggy top as desired. Also known as the physician's car.

Style No. 3

Buggy Top, Raised

Price . . . $900
f.o.b. Bridgeport, Conn.

Data

Seating capacity	Two passengers
Wheels	28 inches in diameter; steel spokes
Tires	2½-inch single-tube pneumatics
Tread	4 feet 6 inches
Weight, empty	690 pounds
Weight, tanks filled	890 pounds
Capacity of fuel tank	5 gallons
Capacity of water tank	21 gallons
Extreme length	7 feet 4 inches
Extreme width	4 feet 10¾ inches
Extreme height	7 feet 5½ inches
Seat	Panel back

Equipment

Rubber bucket, boot, side lamps, cyclometer, gong, full set of tools.

"Locomobile"

Model	Body Style	Notes
4	Locoracer. 1 psgr. spindle seat.	39-3/4 inch tread with standard 78 inch wheelbase. 2 gallon fuel tank and 10 gallon water tank.

Style No. 4

Price . . . $750
f.o.b. Bridgeport, Conn.

Data

Seating capacity	One passenger
Wheels	28 inches in diameter; steel spokes
Tires	2½-inch single-tube pneumatics
Tread	3 feet 3¾ inches
Weight, empty	428 pounds
Weight, tanks filled	540 pounds
Capacity of fuel tank	2 gallons
Capacity of water tank	10 gallons
Extreme length	7 feet 3¼ inches
Extreme width	3 feet 8 inches
Extreme height	4 feet 4¾ inches
Seat	Spindle back

Equipment

Rubber bucket, side lamps, cyclometer, gong, full set of tools.

"Locoracer"

1900

Model	Body Style	Notes
5	Locosurrey. 4 psgr. spindle seats.	6-1/2 gallon gas tank with 21 gallon water tank.

Style No. 5

Price . . . $1200
f.o.b. Bridgeport, Conn.

Data

Seating capacity	Four passengers
Wheels .	28 inches in diameter; steel spokes
Tires . .	2½-inch single-tube pneumatics
Tread	4 feet 6 inches
Weight, empty	800 pounds
Weight, tanks filled	1020 pounds
Capacity of fuel tank	6½ gallons
Capacity of water tank	21 gallons
Extreme length	8 feet 5½ inches
Extreme width . . .	4 feet 10¾ inches
Extreme height	5 feet 3½ inches
Seats	Spindle back

Equipment

Rubber bucket, two rubber blankets, side lamps, cyclometer, gong, full set of tools.

"Locosurrey"

6	Lococycle. 2 psgr.	Two passenger motorcycle with twin small boilers and vertically mounted fuel and water tanks. Only appeared in early 1900 catalog. Not listed later.

Lococycle.

STYLE No. 6.

THE FASTEST PACING MACHINE IN THE WORLD.

1900

Model	Body Style	Notes
02	Stanhope. 2 psgr. panel seat.	Body two inches wider and two inches longer than Style 2. Special finish and upholstery. 5 gallon gas tank and 26 gallon water tank.

Style No. 02

Price . . . $850
f.o.b. Bridgeport, Conn.

Carriage body two inches wider and two inches longer than Style No. 2. Special finish and upholstery

Data

Seating capacity	Two passengers
Wheels	28 inches in diameter; steel spokes
Tires	2½-inch single-tube pneumatics
Tread	4 feet 6 inches
Weight, empty	700 pounds
Weight, tanks filled	950 pounds
Capacity of fuel tank	5 gallons
Capacity of water tank	26 gallons
Extreme length	7 feet 4 inches
Extreme width	4 feet 10¾ inches
Extreme height	5 feet 3¾ inches
Seat	Panel back

Equipment

Rubber bucket, rubber blanket, side lamps, gong, cyclometer, full set of tools.

"Locomobile"

Model	Body Style	Notes
03	Stanhope. 2 psgr. panel seat.	Same as Style 3 but with Victoria top. Apparently, the Company decided to differentiate between models with this top and the buggy top.

Style No. 03

Victoria Top, Raised

Price . . . $900
f.o.b. Bridgeport, Conn.

Data

Seating capacity	Two passengers
Wheels	28 inches in diameter; steel spokes
Tires	2½-inch single-tube pneumatics
Tread	4 feet 6 inches
Weight, empty	700 pounds
Weight, tanks filled	900 pounds
Capacity of fuel tank	5 gallons
Capacity of water tank	21 gallons
Extreme length	7 feet 4 inches
Extreme width	4 feet 10¾ inches
Extreme height	7 feet 4½ inches
Seat	Panel back

Equipment

Rubber bucket, boot, side lamps, cyclometer, gong, full set of tools.

"Locomobile"

1900

Model	Body Style	Notes
003	Stanhope. 2 psgr. panel seat.	Same as Style 02 but with buggy top.

Style No. 003

Buggy Top, Raised

Price . . . $1000

f.o.b. Bridgeport, Conn.

Carriage body two inches wider and two inches longer than Style No. 3. Special finish and upholstery

Data

Seating capacity	Two passengers
Wheels	28 inches in diameter; steel spokes
Tires	2 ½-inch single-tube pneumatics
Tread	4 feet 6 inches
Weight, empty	760 pounds
Weight, tanks filled	1025 pounds
Capacity of fuel tank	5 gallons
Capacity of water tank	26 gallons
Extreme length	7 feet 4 inches
Extreme width	4 feet 10¾ inches
Extreme height	7 feet 5 ½ inches
Seat	Panel back

Equipment

Rubber bucket, boot, side lamps, cyclometer, gong, full set of tools.

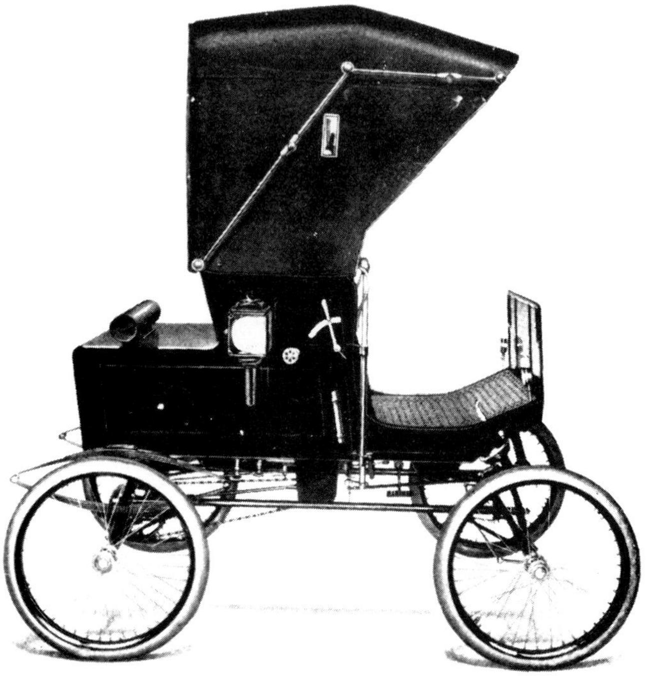

"Locomobile"

Model	Body Style	Notes
0003	Stanhope. 2 psgr. panel seat.	Same as Style 2 but with Victoria top.

Style No. 0003

Victoria Top, Raised

Price . . . $1000

f.o.b. Bridgeport, Conn.

Carriage body two inches wider and two inches longer than Style No. 03. Special finish and upholstery

Data

Seating capacity	Two passengers
Wheels	28 inches in diameter; steel spokes
Tires	2 ½-inch single-tube pneumatics
Tread	4 feet 6 inches
Weight, empty	770 pounds
Weight, tank filled	1025 pounds
Capacity of fuel tank	5 gallons
Capacity of water tank	26 gallons
Extreme length	7 feet 4 inches
Extreme width	4 feet 10¾ inches
Extreme height	7 feet 4 ½ inches
Seat	Panel back

Equipment

Rubber bucket, boot, side lamps, cyclometer, gong, full set of tools.

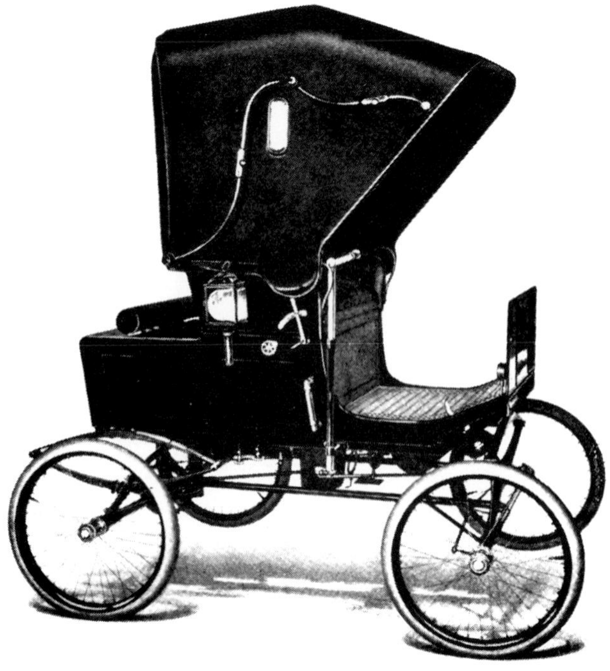

"Locomobile"

1900

Model	Body Style	Notes
05	Surrey. 4 psgr. panel seats.	Body two inches wider and two inches longer than style 5. Special finish and upholstery. 7 gallon gas tank and 26-1/2 gallon water tank.

"Locosurrey"

Style No. 05

Price . . . $1400
f.o.b. Bridgeport, Conn.

Carriage body two inches wider and two inches longer than Style No. 5. Special finish and upholstery

Data

Seating capacity	Four passengers
Wheels . . 28 inches in diameter; steel spokes	
Tires . .	2½-inch single-tube pneumatics
Tread	4 feet 6 inches
Weight, empty	1000 pounds
Weight, tanks filled	1265 pounds
Capacity of fuel tank	7 gallons
Capacity of water tank . . .	26½ gallons
Extreme length	8 feet 5½ inches
Extreme width . . .	4 feet 10¾ inches
Extreme height	5 feet 2 inches
Seats	Panel back

Equipment

Rubber bucket, two rubber blankets, side lamps, cyclometer, gong, full set of tools.

1901

Several changes were introduced this year including a stronger chain tightener, improved differential (spur gear), stronger rear axle housing, double-acting brake, enlarged water tank (Style 2 went from 21 gallons to 26 gallons and Style 02 went from 26 gallons to 32 gallons), feed water heater, 300 pound steam gauge and 100 pound air gauge.

Style	Body	Notes
2	Same as previous	

1901

Model	Body Style	Notes
2-1/2	Stanhope. 2 psgr. Panel seats.	Aluminum rear deck cover. Special finish and upholstery. Same width as Style 2.
02	Same as previous	

Style No. 2½.

Locomobile

PRICE, $900

f. o. b. Bridgeport, Conn.

High-grade dash. Aluminum rear cover. Special finish and upholstery.

Data—

Seating capacity	Two passengers
Wheels	28 inches in diameter; steel spokes
Tires	2½-inch single tube pneumatic
Tread	4 feet 6 inches
Weight, empty	740 pounds
Weight, tanks filled	985 pounds
Capacity of gasolene tank	5 gallons
Capacity of water tank	26 gallons
Extreme length	7 feet 4 inches
Extreme width	4 feet 10¾ inches
Extreme height	5 feet 2¼ inches
Seat	Panel back

Equipment—

Rubber bucket, side lamps, gong, Cyclometer, full set of tools. Locomobile Instruction Book. Pamphlets on care of tires. Pamphlet on Cyclometer.

Model	Body Style	Notes
3	Same as previous	
03	Same as previous	
003	Same as previous	Chain sprockets set up for 3:1 reduction to give more power to heavier car.
0003	Same as previous	Chain sprockets set up for 3:1 reduction to give more power to heavier car.
5	Same as previous	
05	Same as previous	Car now has 16 inch boiler and burner.

1902

Several changes occurred this year also with the introduction of larger cars with more accessories. Several models were apparently tested on the public with little success. These were not catalogued and may never have actually been produced.

Model	Body Style	Notes
2	Same as previous	
02	Same as previous	
03	Same as previous	
0003	Same as previous	
5	Same as previous	Car now has 3:1 gear ratio but still has 14 inch boiler/burner. Two double-acting brakes on differential.
05	Same as previous	Car now has 3:1 gear ratio with 16 inch boiler/burner. Same brakes as above. An option was an extra rear seat with folding footboard that attached to rear similar to dos-a-dos, making this a 6 psgr. car.

1902

Model	Body Style	Notes
Stanhope B	Two psgr. w/panel seats and front box	New design w/16 inch boiler/burner, 10 gallon fuel tank, 36 gallon water holding gas tank steam air and water pumps, Klinger water gauge, cylinder oil pump, mud guards, water tank gauge and steam syphon. Gauge lamp provided. Wheel base now 69 inches instead of standard 58 inches.

Locomobile **Stanhope B**

SPECIFICATIONS

Engine Cylinders	2.5 x 3.5 inches
Boiler and Burner	16 inches diameter
Wheels	28 inches in diameter
Tires	3-inch heavy pneumatics
Wheel Base (axle to axle)	69 inches
Tread (center to center of tires)	4 ft. 6 in.
Capacity of Fuel Tank	10 gallons
Capacity of Water Tank	36 gallons
Weight, empty	900 pounds
Weight, Tanks filled	1,250 pounds
Extreme Width	4 ft. 11 in.
Extreme Length	8 ft. 5 inc.

Equipment includes Steam Air Pump, Steam Water Pump, Klinger Wtaer Gauge, Cylinder Oil Pump, Extra Heavy Differential Gear, new and convenient way of carrying Tools, Large Water Piping, Two Double-acting Band Brakes, Mud Guards, Simplified Piping giving great accessibility to working parts, Lamp for Illuminating Water Gauge at night, Water Tank Gauge, Ejector for filling Water Tanks.

ALSO: Full set of tools, cyclometer, gong, side-lamps, LOCOMOBILE instruction book, pamphlets on care of tires and cyclometer, storm apron, rubber bucket for filling water tank, LOCOMOBILE oil can, which feeds oil under pressure.

Special Stanhope	3 psgr. w/panel seat. Detachable seat is provided behind body for third passenger.	Similar to Stanhope B in equipment but with 9 gallon fuel tank and 35 gallon water tank. Wheel base is 66 inches.

Locomobile **Special Stanhope**

SPECIFICATIONS

Engine Cylinders	2.5 x 3.5 inches
Boiler and Burner	16 inches diameter
Wheels	23 inches in diameter
Tires	3-inch heavy pneumatics
Wheel Base (axle to axle)	66 inches
Tread (center to center of tires)	4 ft. 6 in.
Capacity of Fuel Tank	9 gallons
Capacity of Water Tank	35 gallons
Weight, empty	1,090 pounds
Weight, Tanks filled	1,430 pounds
Extreme Width	5 ft.
Extreme Length	8 ft. 5 inc.

Equipment includes Steam Water Pump, Steam Air Pump, Extra Heavy Differential Gear, new and convenient way of carrying Tools, Large Water Piping, Cylinder Oil Pump, Ejector for filling Water Tank without leaving seat, Klinger Wtaer Gauge, Simplified Piping giving great accessibility to working parts, Gauge showing amount of water in water tank, Lamp for illuminating water gauge at night, Two Double-acting Band Brakes, Mud Guards.

ALSO: Full set of tools, gong, side-lamps, cyclometer, LOCOMOBILE instruction book, pamphlets on care of tires and cyclometer, storm apron, rubber bucket for filling water tank.

1902

Model	Body Style	Notes
Dos-a-Dos	4 psgr. with metal seat rails. Folding rear footboard for rear-facing psgrs. Footboard can fold up and rear seat be removed to allow storage room in back for luggage.	16 inch boiler/burner with 12 gallon fuel tank and 33 gallon water tank. Includes two double-acting brakes but no steam pumps or other extras as in the Stanhope B. No smokestack. Flue exhaust now exits on side of body below rear seat. Fenders supplied.

Locomobile Dos–a–Dos

SPECIFICATIONS

Engine Cylinders	2.5 x 3.5 inches
Boiler and Burner	16 inches diameter
Wheels	28 inches in diameter
Tires	2.5-inch heavy pneumatics
Wheel Base (axle to axle)	58 inches
Tread (center to center of tires)	4 ft. 6 in.
Capacity of Fuel Tank	12 gallons
Capacity of Water Tank	33 gallons
Weight, empty	1,000 pounds
Weight, Tanks filled	1,380 pounds
Extreme Width	4 ft. 11 in.
Extreme Length	7 ft. 10 inc.

Equipment includes Two Double-acting Band Brakes, Simplified Piping giving great accesibilty to working parts, new and convenient way of carrying Tools, Lamp for illuminating water gauge at night, Heavy Differntial Gear.
ALSO: Full set of tools, gong, side-lamps, cyclometer, LOCOMOBILE instruction book, pamphlets on care of tires and cyclometer, storm apron, rubber bucket for filling water tank.

Touring Model A	2 psgr. w/panel seat and front box holding gas tank.	16 inch boiler/burner with 20 gallon gas tank and 46 gallon water tank. Wheel base is 73 inches with 56 inch tread. Equipment included similar to Stanhope B.

Locomobile Touring Model "A"

SPECIFICATIONS
Touring Model A

Boiler and Burner	16 inches diameter
Wheels	28 inches in diameter
Tires	3-inch heavy pneumatics
Wheel Base (axle to axle)	73 inches
Tread (center to center of tires)	4 ft. 8 in.
Capacity of Fuel Tank	20 gallons
Capacity of Water Tank	46 gallons
Weight, empty	1,250 pounds
Weight, Tanks filled	1,730 pounds
Extreme Width	5 ft. 3 in.
Extreme Length	9 ft. 1 inc.

Equipment includes Steam Water Pump, Two Double-acting Band Brakes, Mud Guards, Simplified Piping giving great accessibility to working parts, Heavy Differential Gear, Large Water Piping, new and convenient way of carrying Tools, Lamp for illuminating water gauge at night.
ALSO: Full set of tools, cyclometr, gong, LOCOMOBILE instruction book, pamphlets on care of tires and cyclometer, storm apron, rubber bucket for filling water tank, LOCOMOBILE oil can, which feeds oil under pressure.

1902

Model	Body Style	Notes
Touring Model B	2 psgr. w/panel seats and front box for fuel tank.	20 inch boiler/burner with 22 gallon fuel tank and 50 gallon water tank. Wheelbase is 73 inches with 56 inch tread. Equipment similar to Stanhope B. Mudguards provided.

Locomobile Touring Model "B"

SPECIFICATIONS
Touring Model B

Boiler and Burner	20 inches diameter
Wheels	28 inches in diameter
Tires	3-inch heavy pneumatics
Wheel Base (axle to axle)	73 inches
Tread (center to center of tires)	4 ft. 8 in.
Capacity of Fuel Tank	22 gallons
Capacity of Water Tank	59 gallons
Weight, empty	1,500 pounds
Weight, Tanks filled	2,100 pounds
Extreme Width	5 ft. 4 in.
Extreme Length	9 feet

Equipment includes Steam Water Pump, Klinger Water Gauge, Two Double-acting Band Brakes, Trailer Brake, Mud Guards, Cylinder Oil Pump, Simplified Piping giving great accesibilty to working parts, Heavy Differential Gear, Gauge showing amount of water in Tank, new and convenient way of carrying Tools, Large Water Piping, Lamp for illuminating water gauge at night.

ALSO: Gong, cyclometer, full set of tools, side-lamps, LOCOMOBILE instruction book, pamphlets on care of tires and cyclometer, storm apron, rubber bucket for filling water tank, LOCOMOBILE oil can, which feeds oil under pressure.

Wagonette	6-9 psgr. car with one seat facing forward and two longitudinal seats behind facing each other.	3x4 inch engine with 16 inch boiler/burner combination. 8-1/2 gallon fuel tank and 42 gallon water tank. In addition to the two differential brakes, two hub brakes are provided on the rear wheels. Burner flue now exits on side of body below seats. Locomobile considered this design special enough to acquire a patent on the body.
Loco-delivery	Small enclosed delivery wagon with room for 2 psgrs.	Same running gear as Wagonette with 12-1/2 gallon fuel tank and 42 gallon water tank. Tires are 32x3.

In addition, Locomobile had tested the market with an enclosed depot wagon seating four persons and having the large surrey running gear. No information is available on this model.

1903

Model	Body Style	Notes
Long Wheelbase Runabout	2 psgr w/panel seats and front box for fuel tank	14 inch boiler/burner with 14 gallon fuel tank and 30 gallon water tank. Equipped with steam air and water pumps, Klinger gauge, water syphon, water tank gauge, mudguards, mechanical cylinder lubricator, mechanical engine lubricator. Double-acting brakes on differential.

Stanhope B	Same as previous	18 inch boiler/burner. Pilot light in burner. Engine is now 3x4, encased and running in oil. Roller bearing crossheads. Automatic ratchet to tight packing glands. Wheels are now 30x3-1/2 wood artillery with steering wheel and worm-gear steering. Two seats of brakes, on differential and rear wheel hubs. Brake is now on ratchet to act as parking brake. Throttle is now on steering post. Rest of equipment same as 1902 model.
Dos-a-dos	4 psgr. w/panel front seat and spindle rear seat.	Equipment same as Stanhope B
Touring A	Same as previous	Equipment same as Stanhope B Steering can be by tiller or steering wheel at buyer's choice.

1903

Model	Body Style	Notes
Standard Locomobile	Same as Style 2	14 inch boiler/burner with pilot light. 2-1/2x3-1/2 engine but with superheater and mechanical cylinder lubrication. Mechanical engine lubrication, steam air and water pumps, Klingergauge, steam syphon. Basically the same as the 1899 Style 2 but with improvements.
Special Surrey	4 psgr. w/panel seats. Front box holds fuel tank	18 inch boiler/burner. Equipment same as Stanhope B. Fuel tank is 16 gallon with 35 gallon water tank. Acetylene headlights now supplied as an extra.

Special Surrey, $1,800

SPECIAL SURREY

SPECIFICATIONS. *Boiler,* 18″ Locomobile fire tubular type. Indestructible water gauge with lamp for use in night riding. *Burner,* Improved pattern; very durable and fitted with generator and pilot light. *Engine,* Two double-acting cylinders 3″ x 4″, encased and running in oil. Roller bearing crossheads. Roller bearings throughout. Automatic cylinder oil pump. A lever with ratchet allows operator to loosen or tighten piston glands and valve stem glands instantly, thus avoiding any burning of the fingers. Reverse lever is the regular locomotive type and positive. *Wheels,* Wooden artillery type with 3½″ x 30″ double-tube tires. Mud guards. Rear wheels have roller bearings; front wheels plain bearings. *Brakes,* Two double-acting brakes on rear wheel hubs operated by ratchet lever. Two double-acting brakes on differential. This provides two independent and positive methods of stopping the vehicle. The ratchet brake allows operator to leave carriage standing on a hill. *Steering* is by our regular side-steering device or by wheel, as may be desired. Throttle on steering post. *Tanks,* Water, 35 gallons; fuel, 16 gallons. *Dimensions,* Wheel base 93″; tread, 54″. *Equipment* includes Victor steam air and water pumps. Auxiliary hand water pump. Water tank ejector. Water tank gauge. Side lamps. (Acetylene head lights are extra.) French horn. Storm apron. Cyclometer. Set of tools.

1903

Model	Body Style	Notes
1903 Long Wheelbase Runabout	2 psgr. w/panel seats. Front box holds fuel tank.	16 inch boiler/burner. 2-1/2 x 3-1/2 engine with superheater. Wheels are 28 x 2-1/2 wire but other equipment same as Stanhope B. Fuel tanks holds 14 gallons with 32 gallon water tank.

1903 Long Wheel Base Runabout

SPECIFICATIONS. *Boiler,* 16″ Locomobile fire tubular type. Indestructible water gauge with lamp for use in night riding. *Burner,* Improved pattern; very durable and fitted with generator. *Engine,* Two double-acting cylinders, 2½″ x 3½″, using superheated steam. Automatic cylinder oil pump. Octopus lubricator. *Wheels,* Strong and graceful steel wheels fitted with 2½″ x 28″ thick-walled, single-tube pneumatic tires. Mud guards. *Brakes,* Double-acting brake on differential. *Steering.* Our regular side-steering device. *Tanks,* Water 32 gallons; fuel 14 gallons, carried in front box. *Dimensions,* Wheel base 75″; tread 54″. *Equipment,* Victor steam water and air pumps. Auxiliary hand water pump. Water tank ejector. Two side lamps. Storm apron. French horn. Cyclometer. Set of tools.

Touring B	Same as previous except front box holds extra seat and footboard	Same as 1902 model. Fuel tank holds 17 gallons with 45 gallon water tank. Steering can be by tiller or wheel. 1902 improvements in engine (3x4).
Locosurrey No. 5	Same as previous	Same as 1902 model.
Stanhope B	Same as previous	Same as 1902 model with 16 inch boiler/burner but with 3x4 engine and other 1902 improvements. In other words, two Stanhope B's were made in the same year.

1904

Model	Body Style	Notes

For all practical purposes, the 1904 cars were the same as the 1903 models. Since Locomobile was concentrating it efforts in gas cars, I suspect these were made up of the same parts as the 1902 models. The steam car line was still bringing in the money while the gas car line was getting on its feet.

The models offered were:

Long Wheelbase Runabout	Loco-Surrey No. 5	Stanhope B
Dos-a-Dos	Touring A	Special Surrey
Touring B		

1905

There were reportedly some Locomobiles sold in 1905 but I have not yet seen a catalog of what was offered. I suspect any cars sold were made up of leftover 1903 parts that weren't sold in 1904.

Bibliography

Books

Bacon, John H., *American Steam Car Pioneers*, The Newcomen Society of the United States, Exton, Pennsylvania, 1984.

Beaumont, W. Worby, *Motor Vehicles and Motors*, J. B. Lippincott Co., Philadelphia, 1902.

Clymer, Floyd, *Steam Car Scrapbook*, Bonanza Books, 1945.

Homans, James E., *Self-Propelled Vehicles*, Theo. Audel & Company, 1902.

Lord Montagu of Beaulieu, *Steam Cars 1770-1970*, St. Martin's Press, New York, 1971.

Wortman, Ralph, *A Horseless Carriage Comes to Town*, Oregon Lithoprint, McMinnville, Oregon, 1966.

Magazine Articles

---------, "The Locomobile Steam Carriage," *Scientific American*, January 27, 1900.

Duerksen, Menno, "Locomobile, A Tangled Beginning," *Cars & Parts*, November and December, 1986.

Figge, Chris, "The Chicken Coop Special" *Horseless Carriage Gazette*, May-June, 1977.

Villalon, L. J. Andrew, "The Birth of an Early Automobile Company," *The Bulb Horn*, Veteran Car Club of America, April-June, 1981.

------------, "The Locomobile Company of America, 1899-1905," *The Bulb Horn*, Veteran Car Club of America, July-September and October-December, 1986.

Zordich, James, "The First Automobile into Yosemite," *Horseless Carriage Gazette*, September-October, 1972.